Horticulture Projects

Profiles At A Glance

By

Aditya Kumar Daga
Project & Management Consultant

Project, Marketing And Management Consultant
Astrology, Palmistry & Numerology Adviser
Vastu, Spiritual And Herbal Therapist
Lyricist, Poet, Artist And Composer

"THESE ARE THE YEARS OF HORTICULTURE, AGRICULTURE, TISSU-CULTURE, STEEL & INFRASTRUCTURE AUTOMOBILES, BOILERS & HY-DRAULICS, BIO-TECHNOLOGY, CHEMICALS, HOSPITALITY, PETRO-CHEMICALS, PHARMECEUTICALS, PAPER & PRINTING, SUGAR, TEX-TILE , MEDICAL EQUIPMENTS, INFRASTRUCTURE & SERVICE INDUS-TRIES PROJECTS. "

IN PRESENT INDIA

THIS IS THE REAL TIME "DEMAND PERIOD" FOR THE VALUE ADDED ITEMS WHETHER VALUE ADDED SPECIAL STEEL, NON CARBON STEEL, ALLOY STEEL, CARBON STEEL, STAINLESS STEEL/ ERW/CDW/ PRECISION/SS TUBES & PIPES FABRICATION & GENERAL ENGINEER-ING ITEMS FOR AUTOMOBILES, BOILERS, HYDRAULICS, SUGAR, RE-FINERIES, MEDICAL INDUSTRIES.

SOON THE SUPPLY WILL LAG BEHIND THE DEMAND PROJECTIONS IN ALL THE MAJOR INDUSTRIES SECTOR DUE TO SURPLUS PUCHASING POWER IN THE HANDS OF CONSUMERS AND OVERALL BOOM IN THE PROMINENT INDUSTRIES LIKE POWER, REFINERIES, SUGAR, STEEL, STEEL-VALUE ADDED, PETROCHEMICALS, TEXTILES

<u>NOW THIS IS YOUR TURN TO REEP THE BENEFITS OF THIS BOOMING</u>

<u>ECONOMY.</u>

ARE YOU SERIOUSLY PLANNING FOR :-

*SETTING UP OF:-

ANY NEW PROJECTS DOMESTIC, EXPORT ORIENTED OR 100% EOU,

*EXPANSION OF:-

EXISTING FACILITIES,

*REVIVAL OF :-

ANY OF YOUR CLOSED UNITS,

*REHABILITATION OF :-

ANY OF YOUR SICK UNITS,

*REVALUATION OF :-

YOUR PLANT & MACHINERIES, BUILDING & STRUCTURES;

*DEVELOPING OF :-

DETAILED TEVR , DETAILED MARKET SURVEY REPORTS, DETAILED

PRODUCT SURVEY REPORTS, DETAILED TECHNOLOGY SURVEY RE-

PORTS, DETAILED ANALYTICAL REPORTS, DETAILED POLLUTION CON-

TROL MANAGEMENT REPORTS , DETAILED BUSINESS DEVELOPMENT

REPORTS ; PRESENT BUSINESS PORTFOLIOS

*SEEKING OF :-

COLLABORATIONS, FUND MANAGEMENT, LIAISON & COORDINATION,

STAGE AND/OR TURNKEY PROJECT SERVICES ;

*WE WILL BE :-

DEFINITELY PROVED TO BE ECONOMICAL & MOST RELIABLE ADVI-

SORY PARTNERS PROVIDING ALL THE SERVICES UNDER ONE ROOF

AND READY TO SERVE YOU EITHER ON STAGE BASIS OR TURNKEY BA-

SIS OR RETAINERSHIP CONTRACTUAL BASIS . WE WORK ON ALL IN-

DIA BASIS AND HAVING ENOUGH STAFF STRENGTH & OFFICE

STRENGTH.

SO WE ARE HERE :-

TO ASSIST YOU AT EVERY STAGE SINCE LAST 30 YEARS:-

"MCPL"- A TRUSTED NAME & YOUR CONSISTENT PROFESSIONAL ADVISOR IN THE FOLLOWING & VARIOUS OTHER PROJECTS:-			
INFRASTRUCTURE	CHEMICALS	*TEXTILES*	CONSUMER
Hotels, Resorts, Inns, Spas, Health Clubs Residential Schools, Tech/Mgmt. Institute All NGO Projects, Dairy, Piggy, Poultry Fantasy Spot/ Center, Amusements parks Diagnostic Centers, Nursing Homes Rehabilitation Center, Explorations, Mining Marine Projects, Waste Recycling, Cement Projects	Calcium Carbonates, Diethyl Carbonates, Ethyl Acetate, Acryl ate Latex Gloves, Disposable Syringes Disposable Articles, Detergents Soaps **IRON & STEELS** ERW/CDW Tubes , Rolling Mills, Sponge Iron Plant, Ferro & Other Alloys, Black & Galvanized Tubes	Grey Synthetics, Bleaching/ Dying, Furnishing Fabrics, Spinning Mills, Dyed/Text PY Yarn, Poly Tapes & Levels,EOU Terry Towels, Mattresses, Quilts *CONSTRUCTIONS* Fly Ash Items/ Bricks, Township, Buildings, Hydro & Thermal Power Cement , Handling Eqp.., Geo-Jute Constructions	Glass & China Wares, Electrical Items, Leather Shoes AGRO-BASED IMFL Distilleries, Sugar / Soft Drinks Rice Bran/ Soya Oil,Veg & Refined Oils , Mushroom ,Dried / Dehydrated Vegetables & Fruits, Biotechnology-Oil Floriculture/ Essence, Waste Agro Project

INDIA IS ALWAYS AN IDEAL DESTINATION FOR INDUSTRIES

We provide and prepare exclusive 'Reports' as follows :-

Techno-Economic Viability Reports

Market Survey & Consumer Research Reports,

Product Research Reports,

Varied Technical Reports,Pollution Control Reports,

Environment Assessment Reports,

Civil & Architectural Reports,Revival & Rehabilitations Reports,

Mergers and Amalgamations Reports,

Sensitvity Analysis Reports,

Corporate Progress Reports

Industry Restructure Reports,

Expansion & Diversification Reports,

Business Valuations Reports

Assets Revaluation Reports,

Factory ,Corporate & Office Manuals,

Varied Technical Reports ,

Project Conception to Project Identification Reports,

Locations Search to Land Search,

Project & Vendor Analysis to Evaluation,

Project TEVR's

Institutional Papers & Instutional Documentations

Fund Planning & Appraisal Reports

LISTS OF PROJECTS

PROJECT SYNOPSIS-1   APPLE (Malus Pumila)

PROJECT SYNOPSIS-2   BANANA (*Musa* sp.)

PROJECT SYNOPSIS-3   AONLA (*Emblica officinalis*)

PROJECT SYNOPSIS-4   MANGO (*Mangifera indica* L.)

PROJECT SYNOPSIS-5   GUAVA (*Psidium guajava*)

PROJECT SYNOPSIS-6   ORANGE (MANDARIN)

PROJECT SYNOPSIS-7   PAPAYA (*Carica papaya*)

PROJECT SYNOPSIS-8   SAPOTA (*Achras zapota*)

PROJECT SYNOPSIS-9   PINEAPPLE (*Ananas comosus*)

PROJECT SYNOPSIS-10   LITCHI (*Litchi chinensis*)

PROJECT SYNOPSIS-11   JUJUBE (*Ziziphus mauritiana*)

PROJECT SYNOPSIS-12   POMEGRANATE (*Punica granatum*)

PROJECT SYNOPSIS-13   STRAWBERRY (*Fragaria vesca*)

PROJECT SYNOPSIS-14   KIWI (*Actinidia deliciosa*)

PROJECT SYNOPSIS-15 `   L. GRASS (*Cymbopogan flexuosus*)

PROJECT SYNOPSIS-16   MINT (Labiatae (Lamiaceae)

PROJECT SYNOPSIS-17   PATCHOULI (*Pogostemon patchouli*)

PROJECT SYNOPSIS-18   BUTTON MUSHROOM (*Agaricus spp.*)

PROJECT SYNOPSIS-19   OYSTER MUSHROOM (*Pleurotus* sp.)

PROJECT SYNOPSIS-20   CELERY (*Apium graveolens)*

PROJECT SYNOPSIS-21   GRAPES (*Vitis sp.*)

PROJECT SYNOPSIS -1

APPLE (Malus Pumila)

INTRODUCTION

Apple (Malus pumila) is commercially the most important temperate fruit and is fourth among the most widely produced fruits in the world after banana, orange and grape. China is the largest apple producing country in the world.

OBJECTIVE

The main objective of this report is to present a bankable one-acre model for high quality commercial cultivation of the crop.

BACKGROUN

Origin

Apples originated in the Middle East more than 4000 years ago. Spreading across Europe to France, the fruit arrived in England at around the time of the Norman conquest in 1066.

Area & Production

The area under apple cultivation in India increased by 24% from 1.95 lakh ha. in 1991-92 to 2.42 lakh ha. in 2001-02 although production increased by less than 1% (i.e. from 11 to 12 lakh tones). It is mostly grown in the states of Jammu & Kashmir, Himachal Pradesh, Uttaranchal, Arunachal Pradesh and Nagaland .A one acre plantation of the crop is a highly viable proposition. The cost components of such a model along with the basis for costing are exhibited. A summary is given in the figure below. The project cost works out to Rs. 1.50 lakhs.

COST OF PROJECT			
Project Cost: (Unit – One Acre)			(Amount in
Sl. No.	Component		Proposed Expenditure
1.	Cultivation Expenses		
	(i)	Cost of planting material	2500
	(ii)	Manures & fertilizers	7700
	(iii)	Insecticides & pesticides	3000
	(iv)	Land Preparation	4200
	(v)	Others, if any, (Power)	3600
		Sub Total	21000
2.	Irrigation		
	(i)	Borewell	25000
	(ii)	SIP & Electrical Installation	25000
	(iii)	Others, if any	-
		Sub Total	50000
3.	Cost of Drip/Sprinkler		20000
4.	Infrastructure		
	(i)	Store & Pump House	20000
	(ii)	Labour room	5000
	(iii)	Agriculture Equipments	10000
	(iii)	Others, if any, please specify	
		Sub Total	35000
5.	Land Development		
	(i)	Soil leveling	4000
	(ii)	Digging	-
	(iii)	Fencing	20000
	(iv)	Others, if any, please specify	-
		Sub Total	24000
6.	Land, if newly purchased		-
	Grand Total		1,50,000.00

Inter-cropping

Since the orchard would be start giving yield from 8th year onwards, it is proposed to take up inter-cropping particularly off season vegetables which would cost RS. 10000/- per acre and would yield on average 5 tonnes/acre.

Recurring Production Cost: Recurring production costs are exhibited. The main components are planting material, land preparation, inputs application (FYM, fertilizers, micro-nutrients liming material, plant protection chemicals etc.), power and labour on application of inputs, intercultural and other farm operations.

Besides, provision is included for labour for harvesting and packing/ transportation charges for the produce to the nearest secondary market. The total production cost for a one acre orchard works out to Rs. 174.05 thousand during the first seven years. In the post-operative period i.e. from 8th year onward, the annual recurring cost ranges from Rs.22.00 thousand to Rs.25.30 thousand, at which level it stabilizes.

Returns from the Project: The yield from the plantation is obtained from eighth year onwards. The yield goes up from 3 tonnes per acre to 6 tonnes per acre.   Valued at Rs. 11000 per tonne, the return goes up from Rs. 33 thousand to Rs. 66 thousand

PROJECT SYNOPSIS-2

BANANA (*Musa* sp.)

INTRODUCTION

Banana (*Musa* sp.) is the second most important fruit crop in India next to mango. Its year round availability, affordability, varietal range, taste, nutritive and medicinal value makes it the favourite fruit among all classes of people. It has also good export potential.

Hi-tech cultivation of the crop is an economically viable enterprise leading to increase in productivity, improvement in produce quality and early crop maturity with the produce commanding premium price.

OBJECTIVE

The main objective of this report is to present a bankable model for high quality commercial cultivation of the crop. Efforts need to be made to promote private investment in hi-tech horticulture with micro-propagation, protected cultivation, drip irrigation, integrated nutrient and pest management besides making use of latest post-harvest technologies.

BACKGROUND

Origin

Banana evolved in the humid tropical regions of S.E.Asia with India as one of its centres of origin. Modern edible varieties have evolved from the two species – *Musa acuminata* and *Musa balbisiana* and their natural hybrids, originally found in the rain forests of S.E.Asia. During the seventh century AD its cultivation spread to Egypt and Africa. At present banana is being cultivated throughout the warm tropical regions of the world between 30^0 N and 30^0 S of the equator.

Area & Production

Banana and plantains are grown in about 120 countries. Total annual world production is estimated at 86 million tonnes of fruits. India leads the world in banana production with an annual output of about 14.2 million tonnes. Other leading producers are Brazil, Eucador, China, Phillipines, Indonesia, Costarica, Mexico, Thailand and Colombia.

In India banana ranks first in production and third in area among fruit crops. It accounts for 13% of the total area and 33% of the production of fruits. Production is highest in Maharashtra (3924.1 thousand tones) followed by Tamil Nadu (3543.8 thousand tonnes). Within India, Maharashtra has the highest productivity of 65.70 metric tones /ha. against national average of 30.5 tonnes/ha. The other major banana producing states are Karnataka, Gujarat, Andhra Pradesh and Assam

One acre plantation of the crop is a highly viable proposition. The cost components of such a model along with the basis for costing are exhibited in. A summary is given in the figure below. The project cost works out to Rs. 1.25 lakhs.

Sl. No.	Component		Proposed Expenditure
	COST OF PROJECT		
	Project Cost: (Unit : One Acre)		(Amount in INR Rs.)
1.	Plantation Expenses		
	(i)	Cost of planting material	11,000
	(ii)	Manures & fertilizers	5,000
	(iii)	Insecticides & pesticides	2,000
	(iv)	Cost of Labour	5,000
	(v)	Others, if any, (Power)	2,000
			25,000
2.	Irrigation		
	(i)	Tube-well/submersible pump	40,000
	(ii)	Cost of Pipeline	-
	(iii)	Others, if any, please specify	-
			40,000
3.	Cost of Drip/Sprinkler		25,000
4.	Infrastructure		
	(i)	Store	-
	(ii)	Labour shed & Pump house	10,000
	(iii)	Farm Equipment	1,000
			11,000
5.	Land Development		
	(i)	Soil Leveling	4,000
	(ii)	Digging	-
	(iii)	Fencing	20,000
	(iv)	Others, if any, please specify	-
			24,000
6.	Land, if newly purchased		-
	Grand Total		1,25,000

Inter-cropping

Labour cost has been put at an average of Rs. 70 per man-day. The actual cost will vary from location to location depending upon minimum wage levels or prevailing wage levels for skilled and unskilled labour.

Recurring Production Cost :Recurring production costs are exhibited .The main components are planting material, land preparation, inputs .application (FYM, fertilizers, liming material, plant growth regulators, plant protection chemicals etc.) and labour cost on application of inputs, inter-cultural and other farm operations.

Besides, provision is also included for power charges, protection of the plantation (cost of material for wind protection and polythene bunch covers), labour for harvesting and packing/transportation charges for the produce to the nearest secondary market. The total recurring production cost for a one acre orchard works out as below:

(Rs.Thousand)

Year 1	25.40
Year 2	14.70
Year 3	14.30

Returns from the Project: The yield from the plantation is estimated at 22 tonnes (per acre) comprising 12 tonnes from the main crop and 10 tonnes from the following two ratoon crops. Valued at Rs.6500 per tonne the total realization of the three crops works out to Rs.1.43 lakhs.

PROJECT SYNOPSIS-3

AONLA (*Emblica officinalis*)

INTRODUCTION

Aonla (Emblica officinalis) or Indian gooseberry is indigenous to Indian sub-continent. India ranks first in the world in area and production of this crop. Apart from India naturally growing trees are found in different parts of the world like Sri Lanka, Cuba, Puerto Rico, USA (Hawai & Florida), Iran, Iraq, Pakistan, China, Malaysia, Bhutan, Thailand, Vietnam, Philippines, Trinidad, Panama and Japan.

OBJECTIVE

The main objective of this report is to present a one acre bankable model for high quality commercial cultivation of the crop.

BACKGROUND

Area & Production

Aonla is mostly cultivated in the states of Uttar Pradesh, Maharashtra, Gujarat, Rajasthan, Andhra Pradesh, Karnataka, Tamil Nadu, Himachal Pradesh etc

A one acre plantation of the crop is a highly viable proposition. The cost components of such a model along with the basis for costing and means of financing are exhibited . A summary is given in the figure below. The project cost works out to around Rs. 1.25 lakhs.

COST OF PROJECT

Project Cost: (Unit : One Acre)   (Amount in INR Rs.)

I. No.		Component	Proposed Expenditure
1.		Cultivation Expenses	
	(i)	Cost of planting material	3,200
	(ii)	Input Cost	8,800
	(iii)	Cost of Labour (Land preparation)	2,800
	(iv)	Others, if any, (Power)	3,600
			18,400
2.		Irrigation	
	(i)	Tube-well/submersible pump	50,000
	(ii)	Cost of Pipeline	-
	(iii)	Others, if any, please specify	-
			50,000
3.		Cost of Drip/Sprinkler	20,000
4.		Infrastructure	
	(i)	Labour room & Pump house	7,600
	(ii)	Agriculture Equipments	5,000
			12,600
5.		Land Development	
	(i)	Soil Leveling	4,000
	(ii)	Digging	-
	(iii)	Fencing	20,000
	(iv)	Others, if any, please specify	-
			24,000
6.		Land, if newly purchased	
		Grand Total	1,25,000

Inter-cropping

Labour cost has been put at an average of Rs. 70 per man-day. The actual cost will vary from location to location depending upon minimum wage levels or prevailing wage levels for skilled and unskilled labour.

Recurring Production Cost: Recurring costs in the pre & post-operative period are exhibited. The main components are planting material, land preparation, inputs .application (FYM, fertilizers, micro-nutrients, plant protection chemicals etc.), labour cost on application of inputs and other farm operations, power, harvesting, packing and transportation.

Returns from the Project: In the initial years of development inter-crops will fetch a return of around Rs.24 thousand annually from year 2 to year 5. The yield of the main crop will go up from 4 tonnes in year 1 of commercial production to 8 tonnes in year 5 and will stabilize there-after. The produce has been valued at Rs.700 per quintal in the first year increasing to Rs.1000 per quintal in fourth year and stabilizing thereafter.

PROJECT SYNOPSIS-4

Mango (*Mangifera indica* L.)

INTRODUCTION

Mango (*Mangifera indica* L.) belonging to Family Anacardiaceae is the most important commercially grown fruit crop of the country. It is called the king of fruits. India has the richest collection of mango cultivars.

OBJECTIVE

The main objective of the study is to promote commercial cultivation of the crop by small and middle scale farmers by projecting a one acre bankable model project.

BACKGROUND

Origin

Cultivation of mango is believed to have originated in S.E. Asia. Mango is being cultivated in southern Asia for nearly six thousand years.

Area & Production

India ranks first among world's mango producing countries accounting for about 50% of the world's mango production. Other major mango producing countries include China, Thailand, Mexico, Pakistan, Philippines, Indonesia, Brazil, Nigeria and Egypt. India's share is around 52% of world production i.e. 12 million tonnes as against world's production of 23 million tonnes (2002-03).

An increasing trend has been observed in world mango production averaging 22 million metric tonnes per year. Worldwide production is mostly concentrated in Asia, accounting for 75% followed by South and Northern America with about 10% share.

Area under cultivation and production trends of mangoes in India during 1997-98 to 2001-02 are depicted in graphs 1 & 2. Major producing States are Andhra Pradesh, Bihar, Gujarat, Karnataka, Maharashtra, Orissa, Tamil Nadu, Uttar Pradesh and West Bengal. Other States where mangoes are grown include Madhya Pradesh, Kerala, Haryana, Punjab etc

A one acre plantation of the crop is a highly viable proposition. The cost components of such a model along with the basis for costing are exhibited . A summary is given in the figure below. The project cost works out to around Rs. 1.50 lakhs per acre.

Inter-cropping

Labour cost has been put at an average of Rs. 70 per man-day. The actual cost will vary from location to location depending upon minimum wage levels or prevailing wage levels for skilled and unskilled labour.

Recurring Production Cost: Recurring production costs are exhibited . The main components are planting material, land preparation, inputs .application (FYM, fertilizers, liming material, plant growth regulators, plant protection chemicals etc.) and labour cost on application of inputs, power, inter-cultural and other farm operations, interest on term loan, harvesting, packing and transportation.

Returns from the Project: : In the development stage returns from inter-cropping are estimated at Rs.25,000 annually. The yield from the plantation is estimated at 5 tonnes in the first year of bearing rising to 7 tonnes. The produce has been valued at Rs. 10,000 per tonne in this exercise.

SI. No.	Component		Proposed Expenditure
	COST OF PROJECT		
Project Cost:			(Amount in INR Rs.)
SI. No.	Component		Proposed Expenditure
1.	Cultivation Expenses		
	(i)	Cost of planting material	2,000
	(ii)	Manures & fertilizers	5,000
	(iii)	Insecticides & pesticides	2,000
	(iv)	Cost of Labour	8,400
	(v)	Others, if any, (Power)	3,600
		Subtotal	21,000
2.	Irrigation		
	(i)	Tube-well/submersible pump	45,000
	(ii)	Cost of Pipeline	-
	(iii)	Others, if any, please specify	-
		Subtotal	45,000
3.	Cost of Drip/Sprinkler		25,000
4.	Infrastructure		
	(i)	Store & pump house	15,000
	(ii)	Labour room	5,000
	(iii)	Agriculture Equipments	5,4000
		Subtotal	25,400
5.	Land Development		
	(i)	Soil Leveling	4,000
	(ii)	Fencing	29,600
		Subtotal	33,600
6.	Land, if newly purchased		-
		Grand Total	1,50,000

PROJECT SYNOPSIS-5

Guava (*Psidium guajava*)

INTRODUCTION

Guava (*Psidium guajava*) is one of the important commercial fruits in India. It is the fourth most important fruit after mango, banana and citrus.

OBJECTIVE

The main objective of this report is to present a bankable one-acre model for high quality commercial cultivation of the crop.

BACKGROUND

Origin

Guava is native to tropical America where it occurs wild. It was introduced in India in the seventeen century.

Area & Production

The area under guava cultivation in India increased by 64% from 94 thousand ha. in 1991-92 to 155 thousand ha. in 2001-02 whereas the production increased by 55% from 11 lakh tones to 17 lakh tonnes. Major guava producing states include Uttar Pradesh, Bihar, West Bengal, Maharashtra, Chhattisgarh, Tamil Nadu, Karnataka, Madhya Pradesh, Gujarat and Andhra Pradesh.

A one acre plantation of the crop is a highly viable proposition. The cost components of such a model along with the basis for costing are exhibited . A summary is given in the figure below. The project cost works out to around Rs. 1.25 lakhs / Acre.

COST OF PROJECT

	Project Cost		(Amount in INR Rs.)
SI. No.	Component		Proposed Expenditure
1.	Cultivation Expenses		
	(i)	Cost of planting material	2200
	(ii)	Manures & fertilizers	5000
	(iii)	Insecticides & pesticides	2000
	(iv)	Cost of Labour	7700
	(v)	Others, if any, (Power)	3600
			20,500
2.	Irrigation		
	(i)	Tube-well/submersible pump	40000
	(ii)	Cost of Pipeline	-
	(iii)	Others, if any, please specify	-
			40,000
3.	Cost of Drip/Sprinkler		20000
4.	Infrastructure		
	(i)	Pump house & Labour shed	10,000
	(ii)	Labour room & godown	-
	(iii)	Agriculture Equipments	1,000
	(iv)	Others, if any (Drying platform)	-
			11,000
5.	Land Development		
	(i)	Soil Leveling	4000
	(ii)	Digging	-
	(iii)	Fencing	29500
	(iv)	Others, if any, please specify	-
			33,500
6.	Land, if newly purchased		-
	Grand Total		1,25,000

Inter-cropping

Labour cost has been put at an average of Rs. 70 per man-day. The actual cost will vary from location to location depending upon minimum wage levels or prevailing wage levels for skilled and unskilled labour.

Recurring Production Cost: Recurring costs and returns are exhibited . The main components are planting material, land preparation, inputs .application (FYM, fertilizers, liming material, plant growth regulators, plant protection chemicals etc.), labour cost on application of inputs & inter-cultural and other farm operations, power, harvesting, packing and transportation charges. Inter cropping upto 4th year would also be taken in the project.

The total development cost in the first two years and recurring production cost from year 3 onwards for a one acre orchard works out as below:

	(Rs. Thousand)
Year 1	21.00
Year 2	25.50
Year 3	30.00
Year 4	32.10
Year 5	24.90
Year 6	24.90
Year 7	25.90
Year 8 onwards	27.00

Returns from the Project: : The yield from the plantation is estimated at 3.00 tonnes in the first year of bearing rising to 6 tonnes. The value of the produce accordingly increases from Rs.24.00 thousand to Rs.54.00 thousand, (vide Annexure III). The inter-crop of vegetables is expected to fetch an income of Rs. 30,000 in the second year and Rs. 25000 per annum in the subsequent two years.

PROJECT SYNOPSIS-6

ORANGE (MANDARIN)

INTRODUCTION

Mandarin orange (*Citrus reticulata*) is most common among citrus fruits grown in India. It occupies nearly 40% of the total area under citrus cultivation in India. The most important commercial citrus species in India are the mandarin (*Citrus reticulata*), sweet orange (*Citrus sinensis*) and acid lime (*Citrus aurantifolia*) sharing 41, 23 and 23 % respectively of all citrus fruits produced in the country.

OBJECTIVE

The main objective of this report is to present a bankable one acre model for high quality commercial cultivation of the crop.

BACKGROUND

Area & Production

In India, citrus is grown in 0.62 million ha. area with the total production of 4.79 million tonnes. The area under orange cultivation in India increased by 67% from 1.19 lakh ha. in 1991-92 to 1.99 lakh ha. in 2001-02 and the production increased by 57% (i.e. from 10.58 to 16.60 lakh tonnes). Oranges are mostly grown in the states of Maharashtra, Madhya Pradesh, Tamil Nadu, Assam, Orissa, West Bengal, Rajasthan, Nagaland, Mizoram, Arunachal Pradesh.

A one acre plantation of the crop is a highly viable proposition. The cost components of such a model along with the basis for costing are exhibited. A summary is given in the figure below. The project cost works out to around Rs. 1.75 lakh/Acre.

COST OF PROJECT

Sl. No.		Component	Proposed Expenditure
1.		**Cultivation Expenses**	
	(i)	Cost of planting material	3,000
	(ii)	Manures & fertilizers	6,900
	(iii)	Insecticides & pesticides	4,000
	(iv)	Cost of Labour	12,500
	(v)	Others, if any (Power)	3,600
			30,000
2.		**Irrigation**	
	(i)	Tube-well/submersible pump	45,000
	(ii)	Cost of Pipeline	-
	(iii)	Others, if any, please specify	-
			45,000
3.		**Drip Irrigation System & including fertigation**	26,500
4.		**Infrastructure**	
	(i)	Store & Pump House	30,000
	(ii)	Labour Shed	5,000
	(iii)	Agriculture Equipments	5,000
			40,000
5.		**Land Development**	
	(i)	Land Leveling	4,000
	(ii)	Fencing	29,500
			33,500
6.		Land, if newly purchased	
		Grand Total	1,75000

Inter-cropping

Labour cost has been put at an average of Rs. 70 per man-day. The actual cost will vary from location to location depending upon minimum wage levels or prevailing wage levels for skilled and un-skilled labour.

Production Cost: The main components are planting material, land preparation, inputs .application (FYM, fertilizers, plant growth regulators, plant protection chemicals etc.), labour cost on appli-cation of inputs & inter-cultural and other farm operations, power, harvesting, packing and transportation charges.

Recurring Production Cost: During the development period of five years .

.

Returns from the Project: : The yield from the plantation is esti-mated at 40 qtls. in the first year of bearing rising to 50 qtls. Valued at Rs. 1200/qtl. The return workout to Rs. 48.00 thousand in the first year increasing to Rs. 60.00 thousand in subsequent years. It might be added that during the gestation period there will be return from inter crops estimated at around Rs. 30.00 thousand per annum.

PROJECT SYNOPSIS-7

Papaya (*Carica papaya*)

INTRODUCTION

Papaya (*Carica papaya*)_is a tropical fruit having commercial importance because of its high nutritive and medicinal value. Papaya cultivation had its origin in South Mexico and Costa Rica. Total annual world production is estimated at 6 million tonnes of fruits. India leads the world in papaya production with an annual output of about 3 million tonnes. Other leading producers are Brazil, Mexico, Nigeria, Indonesia, China, Peru, Thailand and Philippines.

OBJECTIVE

The main objective of this report is to present a bankable one acre model for high quality commercial cultivation of the crop.

BACKGROUND

Area & Production

The area under papaya cultivation in India increased by 63% from 45.2 thousand ha. in 1991-92 to 73.7 thousand ha. in 2001-02 and the production increased from 8 lakh tones to 26 lakh tones. Papaya is mostly cultivated in the states of Andhra Pradesh, Karnataka, Gujarat, Orissa, West Bengal, Assam, Kerala, Madhya Pradesh and Maharashtra.

A one acre plantation of the crop is a highly viable proposition. The cost components of such a model along with the basis for costing are exhibited. A summary is given in the figure below. The project cost works out to Rs. 1.25 lakhs.

SI. No.	Component		Proposed Expenditure
COST OF PROJECT			
Project Cost: (Unit – One Acre)			(Amount in INR Rs.)
1.	Cultivation Expenses		
	(i)	Cost of planting material	3400
	(ii)	Manures & fertilizers	6600
	(iii)	Insecticides & pesticides	500
	(iv)	Cost of Labour	8400
	(v)	Others, if any, (Power Charges)	3600
		Total	22,500
2.	Irrigation		
	(i)	Tube-well/submersible pump	45000
	(ii)	Cost of Pipeline	-
	(iii)	Others, if any	-
		Total	45,000
3.	Cost of Drip/Irrigation including fertigation		25,000
4.	Infrastructure		
	(i)	Labour Shed	5000
	(ii)	Farm Implementation	3500
		Total	8,500
5.	Land Development		
	(i)	Land leveling & layout	4000
	(ii)	Fencing	20000
		Total	24,000
6.	Land (if newly purchased)*		
		Grand Total	1,25,000

Inter-cropping

Labour cost has been put at an average of Rs. 70 per man-day. The actual cost will vary from location to location depending upon minimum wage levels or prevailing wage levels for skilled and unskilled labour.

Recurring Production Cost: Recurring production costs are exhibited. The main components are planting material, land preparation, inputs application (FYM, fertilizers, micro-nutrients, plant protection chemicals etc.) and labour cost on application of inputs, inter-cultural and other farm operations.

Besides, provision is included for power charges, protection of the plantation (cost of material for wind protection and polythene bunch covers), labour for harvesting and packing/transportation charges for the produce to the nearest secondary market. The recurring production cost for a one acre orchard works out as below:

(Rs. Thousand)

Year 1	26.50
Year 2	36.69
Year 3	32.71

Returns from the Project: : The yield from the plantation is estimated at 30 tonnes (per acre) the second year and 25 tonnes in the third year. Valued at Rs. 4500 per tonne the total realization works out to Rs. 247.50 thousand over a three year crop cycle.

PROJECT SYNOPSIS-8

Sapota (*Achras zapota*)

INTRODUCTION

Sapota (*Achras zapota*) commonly known as chiku is mainly cultivated in India for its fruit value, while in South-East Mexico, Guatemala and other countries it is commercially grown for the production of chickle which is a gum like substance obtained from latex and is mainly used for preparation of chewing gum.

OBJECTIVE

The main objective of the study is to present a bankable model for commercial cultivation of the crop through adoption of hi-tech practices.

BACKGROUND

Origin

The fruit is a native of Mexico and other tropical countries of South America.

Area & Production

The area under sapota cultivation has increased from 27 thousand ha. in 1991-92 to 52 thousand ha. in 2001-02 and production from 3.96 lakh tonnes to 5.94 lakh tones.

Sapota is mostly grown in the states of Gujarat, Maharashtra, Karnataka, Tamil Nadu, Andhra Pradesh and Kerala.

A one acre plantation of the crop is a highly viable proposition. The cost components of such a model along with the basis for costing are exhibited. A summary is given in the figure below. The project cost works out to Rs. 1.75 lakhs.

COST OF PROJECT

Project Cost: (Unit – One Acre)   (Amount in INR Rs.)

Sl. No.		Component	Proposed Ex-penditure
1.		Cultivation Expenses	
	(i)	Cost of planting material	4000
	(ii)	Manures & fertilizers	11000
	(iii)	Insecticides & pesticides	3250
	(iv)	Cost of Labour	4550
	(v)	Others, if any, (Irrigation/Power Charges)	3600
		Total	26,400
2.		Irrigation	
	(i)	Tube-well/submersible pump	50000
	(ii)	Cost of Pipeline	-
	(iii)	Others, if any	-
		Total	50,000
3.		Drip/Irrigation System	20,000
4.		Infrastructure	
	(i)	Pump House & Store	30000
	(ii)	Labour Shed	5000
	(iii)	Others (Agri. Implements)	10000
		Total	45,000
5.		Land Development	
	(i)	Land leveling & layout	4000
	(ii)	Fencing	29600
		Total	33,600
		Grand Total	1,75,000

Inter-cropping

Labour cost has been put at an average of Rs. 70 per man-day. The actual cost will vary from location to location depending upon minimum wage levels or prevailing wage levels for skilled and unskilled labour.

Recurring Production Cost: Recurring production costs during the gestation period and during post operative period. The main components are planting material, land preparation, inputs application (FYM, chemical fertilizers, plant protection chemicals etc.) and labour cost on application of inputs, inter-cultural and other farm operations.

Besides, provision is also included for power charges, labour for harvesting and packing/transportation charges for the produce to the nearest secondary market.

Returns from the Project: : The yield from the plantation is obtained from fifth year onwards. The production increases from 4.0 tonnes/acre in the fifth year to 8.0 tonnes/acre in the 8th year and stabilizes thereafter. The produce has been valued at Rs. 10.0 per kg. It might be added that during the initial four years return from inter crops will be around Rs. 30,000/- pa.

PROJECT SYNOPSIS-9

Pineapple (*Ananas comosus*)

NTRODUCTION

Pineapple (*Ananas comosus*) is one of the commercially important fruit crops of India. Total annual world production is estimated at 14.6 million tonnes of fruits. India is the fifth largest producer of pineapple with an annual output of about 1.2 million tonnes. Other leading producers are Thailand, Philippines, Brazil, China, Nigeria, Mexico, Indonesia, Colombia and USA.

OBJECTIVE

The main objective of this report is to present a bankable one acre model for high quality commercial cultivation of the crop.

BACKGROUND

Origin

Cultivation of pineapple originated in Brazil and gradually spread to other tropical parts of the world. Pineapple cultivation was introduced to India by Portuguese in 1548 AD.

Area & Production

The area under pineapple cultivation in India increased by 35% from 57 thousand ha. in 1991-92 to 77 thousand ha. in 2001-02 whereas the production increased by 54% from 8 lakh tonnes to 12 lakh tonnes. The states where pineapple is grown include Assam, Meghalaya, Tripura, Manipur, West Bengal, Kerala, Karnataka and Goa. The other states where it is grown on a small scale are Gujarat, Maharashtra, Tamil Nadu, Andhra Pradesh, Orissa, Bihar and Uttar Pradesh.

A One acre plantation of the crop is a viable proposition. The cost components of such a model along with the basis for costing are given . A summary is given in the figure below. The project cost works out to Rs. 1.50 lakhs.

COST OF PROJECTS

Project Cost: (Unit One Acre)   (Amount in INR Rs.)

Sl. No.	Component		Proposed Expenditure
1.	Cultivation Expenses		
	(i)	Cost of planting material	13000
	(ii)	Manures & fertilizers	10000
	(iii)	Insecticides & pesticides	2000
	(iv)	Cost of Labour	4730
	(v)	Others, if any, (Power)	3600
		Total	33,330
2.	Irrigation		
	(i)	Tube-well/submersible pump	40000
	(ii)	Cost of Pipeline	-
	(iii)	Others, if any (Power Charges)	-
		Total	40,000
3.	Cost of Drip/Sprinkler		20,000
4.	Infrastructure		
	(i)	Store & Pump House	15000
	(ii)	Labour room	5000
	(iii)	Agriculture Equipments	5000
	(iii)	Others, if any, (Please specify)	-
		Total	25,000
5.	Land Development		
	(i)	Soil leveling	2100
	(ii)	Digging	-
	(iii)	Fencing	29,580
	(iv)	Others, if any, please specify	-
		Total	31,680
6.	Land, if newly purchased		-
		Grand Total	150,000

Inter-cropping

Labour cost has been put at an average of Rs. 70 per man-day. The actual cost will vary from location to location depending upon minimum wage levels or prevailing wage levels for skilled and unskilled labour.

Recurring Production Cost: Recurring production costs are exhibited. The main components are planting material, inputs application (fertilizers, liming material, micro-nutrients, plant growth regulators, plant protection chemicals etc.), labour cost on application of inputs, inter-cultural and other farm operations and power.

Besides, provision is included for labour on harvesting and packing/ transportation charges for the produce to the nearest secondary market. The total recurring production cost for a one acre orchard works out as below:

(Rs. Thousand)

Year 1	20.33
Year 2	46.30
Year 3	40.30

Returns from the Project: : The yield from the plantation is estimated at 25.0 tonnes (per acre) in the second year declining to 20.0 tonnes in the third year . Valued at Rs. 5000 per tonne the total realization works out to Rs.2.25 lakhs.

PROJECT SYNOPSIS-10

Litchi (*Litchi chinensis*)

INTRODUCTION

Litchi (*Litchi chinensis*) is a delicious juicy fruit of excellent quality. Botanically it belongs to Sapindaceae family. The translucent, flavoured aril or edible flesh of the litchi is popular as a table fruit in India, while in China and Japan it is preferred in dried or canned state.

OBJECTIVE

The main objective of this report is to present a bankable model for high quality commercial cultivation of the crop.

BACKGROUND

Origin

The origin of litchi is from southern China, particularly the provinces of Kwangtung and Fukien. The spread of litchi to other parts of the world was rather slow probably due to its soil, climatic requirements and short life span of its seed. Litchi reached India through Myanmar and North East region during the 18th Century.

Area & Production

India is the second largest producer of litchi in the World after China. Other major producing countries are Thailand, Australia, South Africa, Madagascar and Florida in US. Among fruit crops, litchi ranks seventh in area and ninth in production but is sixth in terms of value in India. The national average productivity of litchi is 6.1 t/ha, which is much lower than the realizable yield of the crop under well managed condition. The average productivity of litchi in Bihar is 8.0 tonnes/ha. and in West Bengal it is 10.5 tonnes/ha. In other states the productivity is much lower, the lowest of 1.0 t / ha in Uttaranchal. A one acre plantation of the crop is a highly viable proposition. The cost components of such a model along with the basis for costing are exhibited. A summary is given in the figure below. The project cost works out to Rs. 1.50 lakh per acre.

COST OF PROJECT			
Project Cost:			(Amount in INR Rs.)
Sl. No.		Component	Proposed Expenditure
1.		Cultivation Expenses	
	(i)	Cost of planting material	2400
	(ii)	Manures & fertilizers	5000
	(iii)	Insecticides & pesticides	3000
	(iv)	Cost of Labour	5600
	(v)	Others, if any (Power)	3600
		Total	19600
2.		Irrigation	
	(i)	Tube-well/submersible pump	50000
	(ii)	Cost of Pipeline	-
	(iii)	Others, if any, please specify	-
		Total	50000
3.		Cost of Drip/Sprinkler	20000
4.		Infrastructure	
	(i)	Store & Pump House	10000
	(ii)	Labour room	6800
	(iii)	Agriculture Equipments	10000
	(iii)	Others, if any, (Drying platform)	-
		Total	26800
5.		Land Development	
	(i)	Soil leveling	4000
	(ii)	Digging	-
	(iii)	Fencing	29600
	(iv)	Others, if any, please specify	-
		Total	33600
6.		Land, if newly purchased	-
		Grand Total	1,50,000

Inter-cropping

Labour cost has been put at an average of Rs. 70 per man-day. The actual cost will vary from location to location depending upon minimum wage levels or prevailing wage levels for skilled and unskilled labour.

Recurring Production Cost: Recurring production costs are exhibited. The main components are planting material (80 plants/per acre at 7x7m spacing), land preparation, inputs application (FYM, fertilizers, liming material, plant protection chemicals etc.) and labour cost on application of inputs, inter-cultural and other farm operations.

Besides, provision is also included for power charges, labour for harvesting and packing/transportation charges for the produce to the nearest secondary market.

Inter-cropping with vegetables from year 2 to year 5 has been taken into consideration for economic viability of the project

Returns from the Project: : The yield from the plantation is estimated to go up from 2.0 tonnes in year 5 to 6.0 tonnes in the year 9 at which it levels off. The produce has been valued at Rs. 15,000 per tonne.

PROJECT SYNOPSIS-11

Ber or Indian jujube (*Ziziphus mauritiana*)

INTRODUCTION

Ber or Indian jujube (*Ziziphus mauritiana*) is one of the hardy minor fruit crops suitable for cultivation in arid conditions. It is native to India.

OBJECTIVE

The main objective of this report is to present a one acre bankable model for high quality commercial cultivation of the crop.

BACKGROUND

Area & Production

The major ber-growing states are Haryana, Punjab, Uttar Pradesh, Rajasthan, Gujarat, Madhya Pradesh, Bihar, Maharashtra, Andhra Pradesh and Tamil Nadu.

A one acre plantation of the crop is a viable proposition. The major cost components of such a model are given in the table below The project cost works out to Rs. 1.00 lakhs

COST OF PROJECT

Project Cost: (Amount in INR Rs.)

Sl. No.	Component		Proposed Expenditure
1.	Cultivation Expenses		
	(i)	Cost of planting material (6x6m)	2400
	(ii)	Manures & fertilizers	3000
	(iii)	Insecticides & pesticides	2000
	(iv)	Cost of Labour	4000
	(v)	Others, if any, (Power)	3600
		Sub Total	15000
2.	Irrigation		
	(i)	Tube-well/submersible pump	35000
	(ii)	Cost of Pipeline	-
	(iii)	Others, if any	-
		Sub Total	35000
3.	Cost of Drip/Sprinkler		15000
4.	Infrastructure		
	(i)	Store & Pump House	10000
	(iii)	Agriculture Implements	1000
	(iii)	Others, if any, please specify	-
		Sub Total	11000
5.	Land Development		
	(i)	Soil leveling	4000.00
	(ii)	Digging	-
	(iii)	Fencing	20000
	(iv)	Others, if any, please specify	-
		Sub Total	24000
	Grand Total		1,00,000

Inter-cropping

Labour cost has been put at an average of Rs. 70 per man-day. The actual cost will vary from location to location depending upon minimum wage levels or prevailing wage levels for skilled and unskilled labour.

Since the orchard would be start giving yield from 3rd year onwards, it is proposed to take up inter-cropping in the initial years particularly off season vegetables which would cost Rs. 10000/- per acre and would yield on average 6 tonnes/acre. Valued at Rs. 30,000 per annum.

Recurring Production Cost: Recurring production costs are exhibited. The main components are planting material, land preparation, inputs application (FYM, fertilizers, micro-nutrients liming material, plant protection chemicals etc.), power and labour on application of inputs, inter-cultural and other farm operations.

Returns from the Project: : The yield from the plantation is obtained from third year onwards.   The yield per tree increases from 3 tonnes per acre in the 3rd year to 7 tonnes per year. Valued at Rs. 8 per kg., the return accordingly goes up from Rs. 24 thousand to Rs. 56 thousand .

PROJECT SYNOPSIS-12

Pomegranate (*Punica granatum*)

NTRODUCTION

Pomegranate (*Punica granatum*) is one of the commercially important fruit crops of India. It is native to Iran (Persia).

OBJECTIVE

The main objective of this report is to present a one acre bankable model for high quality commercial cultivation of the crop.

BACKGROUND

Area & Production

Pomegranate is cultivated commercially only in Maharashtra. Small scale plantations are also seen in Gujarat, Rajasthan, Karnataka, Tamil Nadu , Andhra Pradesh, Uttar Pradesh, Punjab and Haryana.

A one acre plantation of the crop is a viable proposition. Project cost of the model, along with the basis for costing are exhibited. A summary of the project cost is given in the table below. The project cost works out to Rs. 1.75 lakhs.

COST OF PROJECT

Project Cost: (Amount in INR

Sl. No.		Component	Proposed Expenditure
1.		Cultivation Expenses	
	(i)	Cost of planting material	4000
	(ii)	Manures & fertilizers	11000
	(iii)	Insecticides & pesticides	4000
	(iv)	Cost of Labour	8800
	(v)	Others, if any, (Power)	3600
		Sub Total	31400
2.		Irrigation	
	(i)	Tube-well/submersible pump	45000
	(ii)	Cost of Pipeline	-
	(iii)	Others, if any	-
		Sub Total	45000
3.		Cost of Drip/Sprinkler	20000
4.		Infrastructure	
	(i)	Store & Pump House	30000
	(ii)	Labour shed	5000
	(iii)	Agriculture Equipments & Implements	10000
	(iii)	Others, if any, please specify	-
		Sub Total	45000
5.		Land Development	
	(i)	Soil leveling	4000
	(ii)	Digging	-
	(iii)	Fencing	29600
	(iv)	Others, if any, please specify	-
		Sub Total	33600
6.		Land, if newly purchased	-
		Grand Total	1,75,000

Inter-cropping

Labour cost has been put at an average of Rs. 70 per man-day. The actual cost will vary from location to location depending upon minimum wage levels or prevailing wage levels for skilled and unskilled labour.

Since the orchard would be start giving yield from 5th year onwards, it is proposed to take up inter-cropping particularly off season vegetables which would cost Rs. 10000/- per acre and would yield on average 6 tonnes/acre valued at Rs. 30000.

Recurring Production Cost: Recurring production costs are exhibited . The main components are planting material, land preparation, inputs application (FYM, fertilizers, micro-nutrients liming material, plant protection chemicals etc.), power and labour on application of inputs, inter-cultural and other farm operations.

Returns from the Project: : The yield from the plantation is obtained from 5th year onwards. The yield goes up from 4.0 tonnes per acre in the 5th year to 7 tonnes per acre in the 8th year onwards. Valued at Rs. 15,000 per tonne the return goes up from Rs. 0.60 lakhs to Rs. 1.05 lakhs .

PROJECT SYNOPSIS-13

Strawberry (*Fragaria vesca*)

INTRODUCTION

Strawberry (*Fragaria vesca*) is an important fruit crop of India and its commercial production is possible in temperate and sub-tropical areas of the country.

OBJECTIVE

The main objective of this report is to present a bankable one-acre model for high quality commercial cultivation of the crop.

BACKGROUND

Area & Production

Strawberry is cultivated in Himachal Pradesh, Uttar Pradesh, Maharashtra, West Bengal, Delhi, Haryana, Punjab and Rajasthan. Sub-tropical areas in Jammu have also the potential to grow the crop under irrigated condition.

Estimates of area and production of the crop are not available.

A one acre plantation of the crop is a viable proposition. Project cost of the model, along with the basis for costing are exhibited . A summary of the project cost is given in the table below.

		COST OF PROJECT	
Project Cost:			(Amount in INR Rs.)
Sl. No.		Component	Proposed Expenditure
1.		Cultivation Expenses	
	(i)	Cost of planting material	200000
	(ii)	Fertilizers & Pestsicides	11000
	(iii)	Mulching	12400
	(iv)	Cost of Labour	14400
	(v)	Others, if any, (Power)	3600
		Sub Total	241000
2.		Irrigation	
	(i)	Tube-well/submersible pump	50000
	(ii)	Cost of Pipeline	-
	(iii)	Others, if any	-
		Sub Total	50000
3.		Cost of Drip (Turboline) with Fertigation	40000
4.		Infrastructure	
	(i)	Store & Pump House	20000
	(ii)	Labour room	10000
	(iii)	Agriculture Equipments & Implements	5000
	(iii)	Others, if any, please specify	-
		Sub Total	35000
5.		Land Development	
	(i)	Soil leveling	4000
	(ii)	Digging	-
	(iii)	Fencing	29600
	(iv)	Others, if any, please specify	-
		Sub Total	33600
		Grand Total	4,00,000

Inter-cropping

Labour cost has been put at an average of Rs. 70 per man-day. The actual cost will vary from location to location depending upon minimum wage levels or prevailing wage levels for skilled and unskilled labour.

Recurring Production Cost: Recurring production costs are exhibited . The main components are planting material, land preparation, inputs application (FYM, fertilizers, micro-nutrients liming material, plant protection chemicals etc.), power and labour on application of inputs, inter-cultural and other farm operations.

Returns from the Project: : The strawberry is short duration crop. The crop planted in September-October starts going yield in May-June. It continues to give yield upto 3rd year thereafter it needs re-planted. Average yield of strawberry is 8 tonnes/acre with good management. The average sale rate is Rs. 40,000 per tonne. Thus gross return works out to Rs.3.20 lakhs per acre/annum.

PROJECT SYNOPSIS-14

Kiwi or Chinese gooseberry (*Actinidia deliciosa*)

INTRODUCTION

Kiwi or Chinese gooseberry (*Actinidia deliciosa*) is grown widely in New Zealand, Italy, USA, China, Japan, Australia, France, Chile and Spain.

OBJECTIVE

The main objective of this report is to present a bankable one acre model for high quality commercial cultivation of the crop.

BACKGROUND

Area & Production

Kiwi is mostly grown in the mid hills of Himachal Pradesh, Uttar Pradesh, J & K, Sikkim, Meghalaya, Arunachal Pradesh and Kerala. Having been very newly introduced in the country estimates of area and production have not yet become available.

A one acre plantation of the crop is a viable proposition. The major cost components of such a model are given in the table below: The project cost works out to Rs. 2.50 lakhs .

COST OF PROJECT

Project Cost: (Amount in INR Rs.)

Sl. No.	Component		Proposed Expenditure
1.	Cultivation Expenses		
	(i)	Cost of planting material (4x5m)	4000
	(ii)	Manures & fertilizers	6000
	(iii)	Insecticides & pesticides	2000
	(iv)	Cost of Labour	5800
	(v)	Others, if any, (Power)	3600
		Sub Total	21400
2.	Irrigation		
	(i)	Tube-well/submersible pump	50000
	(ii)	Cost of Pipeline	-
	(iii)	Others, if any	-
		Sub Total	50000
3.	Cost of Drip/Sprinkler		20000
4.	Infrastructure		
	(i)	Store & Pump House	20000
	(ii)	Labour room	10000
	(ii)	Agriculture Equipments/ Implements	5000
	(iii)	Others, if any (Training Structure)	90000
		Sub Total	125000
5.	Land Development		
	(i)	Soil leveling	4000
	(ii)	Digging	-
	(iii)	Fencing	29600
	(iv)	Others, if any, please specify	-
		Sub Total	33600
	Grand Total		2,50,000

Inter-cropping

Since the orchard would be start giving yield from 4th year onwards, it is proposed to take up inter-cropping in the initial years particularly of vegetables which would cost RS. 10000/- per acre and would yield on average 6 tonnes/acre. Valued at Rs.30.00 thousand.

Recurring Production Cost: Recurring production costs are exhibited. The main components are planting material, land preparation, inputs application (FYM, fertilizers, micro-nutrients liming material, plant protection chemicals etc.), power and labour on application of inputs, inter-cultural and other farm operations.

Returns from the Project: : The yield from the plantation is obtained from fourth year onwards. The yield per acre increases from 5 tonnes in 4th year to 8 tonnes in 7th year. Valued at Rs. 18000 per tonne the return goes up from Rs. 90 thousand to Rs. 144 thousand.

PROJECT SYNOPSIS-15

Lemon grass (*Cymbopogan flexuosus*)

INTRODUCTION

Lemon grass (*Cymbopogan flexuosus*) is a native aromatic tall sedge (family: Poaceae) which grows in many parts of tropical and sub-tropical South East Asia and Africa. In India, it is cultivated along Western Ghats (Maharashtra, Kerala), Karnataka and Tamil Nadu states besides foot-hills of Arunachal Pradesh and Sikkim. It was introduced in India about a century back and is now commercially cultivated in these States.

OBJECTIVE

The main objective of this report is to present a bankable one acre model for high quality commercial cultivation of the crop.

BACKGROUND

Origin

Most of the species of lemon grass are native to South Asia, Southeast Asia and Australia. The so called East Indian lemon grass (*Cymbopogon flexuosus*) , also known as *Malabar or Cochin grass is native* to India , Sri Lanka, Burma and Thailand ; for the related West Indian lemon grass (*C. citratus*), a Malesian origin is generally assumed. Both the species are today cultivated throughout tropical Asia.

Area & Production

At present, India grows this crop in 3,000 ha area, largely in states of Kerala, Karnataka, U.P. and Assam and the annual production ranges between 300-350 t/annum.

COST OF PROJECT

Project Cost: (Unit – One Acre)   (Amount in INR Rs.)

Sl. No.		Component	Proposed Expenditure
1.		Cultivation Expenses	
	(i)	Cost of planting material	200
	(ii)	Input Cost	4500
	(iii)	Insecticides & pesticides	2000
	(iv)	Cost of Labour	7000
	(v)	Others, if any, please specify (Power)	3600
		Total	17300
2.		Irrigation	
	(i)	Tube-well/submersible pump	32000
	(ii)	Pump & Electrical Installation	25000
	(iii)	Others, if any	-
		Total	57000
3.		Cost of Drip/Sprinkler	-
4.		Infrastructure	
	(i)	Store & Pump House	20000
	(ii)	Labour room	-
	(iii)	Agriculture Equipments	6100
	(iii)	Others, if any, (Drying platform)	16000
		Total	42100
5.		Land Development	
	(i)	Soil leveling	4000
	(ii)	Digging	-
	(iii)	Fencing	29600
	(iv)	Others, if any, please specify	-
		Total	33600
6.		Land, if newly purchased	-
		Grand Total	1,50,000

Inter-cropping

Labour cost has been put at an average of Rs. 70 per man-day. The actual cost will vary from location to location depending upon minimum wage levels or prevailing wage levels for skilled and unskilled labour.

Recurring Production Cost: Recurring production costs are exhibited . The main components are planting material, purchase of inputs, power and labour cost on land preparation, application of inputs, inter-cultural and other farm operations.

Besides, provision is also included for processing (extraction of oil) and marketing. The total annual recurring cost for a one acre farm works out to Rs. 4.5 thousand (approx.)

Returns from the Project: : The yield from the farm is estimated at 100 kg. of oil. Valued at Rs. 500 per kg. the total realization of the three crops works out to Rs. 1.50 lakhs .

PROJECT SYNOPSIS-16

MINT (MENTHA) – (Labiatae (Lamiaceae)

INTRODUCTION

Mints belong to the genus *Mentha*, in the family Labiatae (Lamiaceae) which includes other commonly grown essential oil-yielding plants such as basil, sage, rosemary, marjoram, lavender, pennyroyal and thyme. Within the genus *Mentha* there are several commercially grown species, varying in their major chemical content, aroma and end use. Their oils and derived aroma compounds are traded world-wide.

OBJECTIVE

The main objective of this report is to present a bankable one acre model for high quality commercial cultivation of the crop.

BACKGROUND

Origin

The cultivation of Japanese or corn mint originated from Brazil and China. Subsequently, China and India overtook Brazil and more recently India has taken the leading position in cultivation of this essential oil yielding plant.

Area & Production

At present, Japnese mint is cultivated in India on about 60,000 ha. of land with estimated production of 12,000 tonnes of mint oil which accounts for about 75% of total menthol mint production in the world. The economics of a one acre plantation of the crop is a highly viable proposition. The cost components of such a model along with the basis for costing are exhibited. A summary is given in the figure below. The project cost works out to Rs. 1.59 lakhs.

SL. No.		Component	Proposed Expenditure
		COST OF PROJECT	
		Project Cost: (Unit – One Acre)	(Amount in INR Rs.)
1.		Cultivation Expenses	
	(i)	Cost of planting material	1000
	(ii)	Input Cost	6500
	(iii	Cost of Labour	9300
	(iv)	Others, if any, (Power/Irrigation)	3600
		Total	20,400
2.		Irrigation	
	(i)	Tube-well/submersible pump	57,000
	(ii)	Cost of Pipeline	
	(iii)	Others, if any, please specify	
		Total	57,000
3.		Cost of Drip/Sprinkler	-
4.		Infrastructure	
	(i)	Store & pump house	30000
	(ii)	Distillation unit	50000
	(iii)	Agriculture Equipments	10000
	(iv)	Others, if any (Drying platform)	4000
		Total	49,000
5.		Land Development	
	(i)	Soil Levelling	4000
	(ii)	Digging	-
	(iii)	Fencing	29600
	(iv)	Others, if any, please specify	-
		Total	33,600
6.		Land, if newly purchased	
		Grand Total	1,60,000

Inter-cropping

Labour cost has been put at an average of Rs. 70 per man-day. The actual cost will vary from location to location depending upon minimum wage levels or prevailing wage levels for skilled and unskilled labour.

Recurring Production Cost: Recurring production costs are exhibited.   The main components are planting material, land preparation, purchase of inputs, labour cost on application of inputs, inter-cultural & other farm operations and power.

Besides, provision is also included for processing (extraction of oil) and marketing.   The total recurring production cost for a one acre farm works out to Rs. 25.75 thousand per annum.

Returns from the Project: : The per annum yield from the plantation is estimated at 100 kg in terms of oil.   Valued at Rs. 600 per kg. the annual return is Rs. 60 thousand.   One acre mint would also given 25 qtl. of suckers which @ Rs. 400/qtl. would also given income of Rs. 10,000/-.   Thus the total income would be Rs. 70,000/- per acre.

PROJECT SYNOPSIS-17

Patchouli (*Pogostemon patchouli*)

INTRODUCTION

Patchouli (*Pogostemon patchouli*) is a branched, erect, perennial aromatic herb with quardiangular stems (family: Lamiaceae). It is considered to be a native of Phillippines and Malaysia. The leaves are covered with trichomes all over the epidermis, which contains the essential oil. The oil is obtained by steam distillation of shade dried leaves.

The herb is grown extensively in tropical climate of Indonesia, Malaysia, Singapur, China and Brazil, preferably under partial shade.

OBJECTIVE

The main objective of this report is to present a bankable one-acre model for high quality commercial cultivation of the crop.

BACKGROUND

Area & Production

The world production of oil is around 800 t/annum. Java produces 2/3 of this quantity followed by China and Malaysia. Cultivation in India has been meager but is picking up in the last 5 years and is around 600 ha, producing 20 tonnes of oil per annum. It is cultivated in coastal regions of Tamil Nadu, Karnataka, Assam and West Bengal.

The economics of a one acre plantation of the crop is a highly viable proposition. The cost components of such a model along with the basis for costing are exhibited . A summary is given in the figure below. The project cost works out to Rs.2.20 lakhs.

COST OF PROJECT

Project Cost: (Amount in INR Rs.)

Sl. No.		Component	Proposed Expenditure
1.		Cultivation Expenses	
	(i)	Cost of planting material	12000
	(ii)	Manures & fertilizers	8500
	(iii)	Insecticides & pesticides	2600
	(iv)	Cost of Labour	6700
	(v)	Others, if any, (Power)	3600
		Total	33,400
2.		Irrigation	
	(i)	Tube-well/submersible pump	60000
	(ii)	Cost of Pipeline	-
	(iii)	Others, if any, please specify	-
		Total	60,000
3.		Cost of Drip/Sprinkler	20,000
4.		Infrastructure	
	(i)	Pump house and labour shed	22500
	(ii)	Distillation unit & Store	22500
	(iii)	Agriculture Equipments & Machinery	20000
	(iv)	Others, if any (Drying platform)	8000
		Total	73,000
5.		Land Development	
	(i)	Soil Leveling	4000
	(ii)	Digging	-
	(iii)	Fencing	29600
	(iv)	Others, if any, please specify	-
		Total	33,600
6.		Land, if newly purchased	-
		Grand Total	2,20,000

Inter-cropping

Labour cost has been put at an average of Rs. 70 per man-day. The actual cost will vary from location to location depending upon minimum wage levels or prevailing wage levels for skilled and unskilled labour.

Recurring Production Cost: Recurring production costs are exhibited . The main components are planting material, land preparation, purchase of inputs, labour cost on application of inputs, inter-cultural & other farm operations and power.

Besides, provision is also included for processing (extraction of oil) and marketing. The total recurring production cost for a one acre farm increase from Rs.21.14 thousand in the first year to Rs.23.98 thousand in the second and third year.

Returns from the Project: : The per annum yield from the plantation is estimated at 1280 kg in terms of leaves in the first year and 1500 kgs. In second and third year. The value of the resultant produce, in terms of oil, is estimated at Rs.72 thousand in first year increasing Rs.84 thousand per annum in subsequent years.

PROJECT SYNOPSIS-18

Button Mushroom (*Agaricus spp.*)

INTRODUCTION

Button Mushroom (*Agaricus spp.*) is the most popular mushroom variety grown and consumed the world over. In India, its production earlier was limited to the winter season, but with technology development, these are produced almost throughout the year in small, medium and large farms, adopting different levels of technology. The species being grown in most farms is the white button mushroom (*Agaricus bisporus*) belonging to Class Basidiomycetes and Family Agaricaceae.

OBJECTIVE

The main objective of the exercise is to present a small scale viable bankable model production unit through adoption of appropriate technology, utilization of resources and suitable market strategy.

BACKGROUND

Origin

Cultivation of button mushrooms (*A. bisporus*) started in the sixteenth century. However, on a commercial scale, the cultivation was initiated in Europe around 17th Century. Many farms for commercial scale, the cultivation was initiated in Europe around 17th Century. Many farms for production of button mushrooms were established and this variety still dominates the world production and consumption. India, with its diverse agroclimate conditions and abundance of agricultural wastes, has been producing mushrooms, mainly for the domestic market, for more than four decades. Commercial production picked up in the nineties and several hi-tech export oriented farms were set up with foreign technology collaborations. But major share of mushroom production is still on small farms.

Area & Production

Large scale white button mushroom production is centred in Europe (mainly western part), North America (USA, Canada) and S.E. Asia (China, Korea, Indonesia, Taiwan and India). The national annual production of mushrooms is estimated to be around 50,000 tonnes with 85 percent of this production being of button mushrooms

The minimum viable production unit will require a land site of 1.5 acres. The cost components of this model along with the basis for costing are exhibited . A summary is given in the figure below. Inclusive of contingencies, the project cost works out to Rs.107 lakhs as below.

Inter-cropping

Returns from the Project: : To the gives data on production cost and profitability. The yield from the Unit is estimated at 200 tonnes per annum. Valued at Rs.24,000 per tonne, the annual gross return would come to Rs.48 lakhs.

COST OF PROJECT

Project Cost: (Rs. In Lakhs)

Project cost	Amount
Land & Site Development	5.15
Building	44.96
Plant & Machinery	47.00
Misc. Fixed Assets	0.75
Contingency	4.88
Pre-Operative Cost	4.25
Total	106.99

PROJECT SYNOPSIS-19

Oyster mushroom (*Pleurotus* sp.)

INTRODUCTION

Oyster mushroom (*Pleurotus* sp.) belonging to Class Basidiomycetes and Family Agaricaceae is popularly known as 'dhingri' in India and grows naturally in the temperate and tropical forests on dead and decaying wooden logs or sometimes on dying trunks of deciduous or coniferous woods. It may also grow on decaying organic matter. The fruit bodies of this mushroom are distinctly shell or spatula shaped with different shades of white, cream, grey, yellow, pink or light brown depending upon the species.

It is one of the most suitable fungal organisms for producing protein rich food from various agro-wastes or forest wastes without composting.

OBJECTIVE

The main objective of the exercise is to present a small scale viable bankable model production unit using modern technology.

BACKGROUND

Origin

Cultivation of a sp. of oyster mushroom (*Pleurotus ostreatus*) was initiated on experimental basis in Germany by Flack during the year 1917 on tree stumps and wood logs. Growing technology was perfected in USA by Block, Tsao and Hau. Cultivation of different varieties of oyster mushroom was initiated in India in the early sixties. Commercial cultivation began in mid-seventies.

Area & Production

Oyster mushrooms are the third largest cultivated mushroom. China, the world leader in Oyster production, contributes nearly 85% of the total world production of about a million tonnes. The other countries producing oyster mushrooms include Korea, Japan, Italy, Taiwan, Thailand and Phillipines. The present production of this crop in India is only around 1500 tonnes due to low domestic demand. Another inhibiting factor is that export demand orders are large and can be met only if a linkage is developed between producer, cooperatives and exporters.

cost components of such a model along with the basis for costing are exhibited in *Annexures I*. A summary is given in the figure below. Inclusive of 5% contingencies, the project cost works out to around Rs.50 thousand.

Inter-cropping

Recurring Production Cost(Rs. 6.83 thousand): Recurring production costs are brought out . The main components are raw material like wheat straw or rice bran, chemicals, cost of power & water and packaging material etc. Labour costs have been computed at Rs. 80 per man-day. These can, however, vary from location to location depending upon prevailing wage level or minimum statutory wages fixed. Recurring costs work out to Rs. 6.83 thousand per annum.

Returns from the Project: : The yield from this unit would be 400 kgs. per annum. Valued at Rs. 40 per kg. the gross return would be Rs. 16 thousand per annum.

COST OF PROJECT

Project Cost	(Rs. In Lacs)
Project Cost	Amount
Land & Site Development	21.47
Building	15.00
Plant & Machinery	11.90
Contingency	1.42
Total	49.79

PROJECT SYNOPSIS-20
CELERY (*Apium graveolens)*

INTRODUCTION

The celery plant (*Apium graveolens)* belonging to family Apiaceae is a hardy biennial, occasionally annual, widely cultivated for its fleshy leaf-stalk used as a vegetable and seeds which yield essential oil. The seed contains 2 – 3% essential oil and 17 – 18% fatty oil. The essential oil has d-selenene, sedlanolide and sedanoic acid anhydride contributing to its flavour and 60% of d-limonene. The crop is grown as a winter an-nual for its seed and seed oil, used for flavouring tinned food and sauces. It is also used in pickles. The seed has carminative and nerve stimulant properties; it is used as a neuro-tonic in domestic medicine. The leaves are used in salad and also cooked as vegetable.

OBJECTIVE

The crop area at present is 5000ha and the produce is exported as seed mainly to USA.

 The objective is to grow commercial crop of celery for production of seed oil and to raise it for export. Price realization for exported seed is low. Efforts need to be made to produce value added seed oil and export it.

BACKGROUND

Origin

Celery (*Apium graveolens linn.)* is a Mediterranean herb, brought into cultivation in France and England during sixteenth century. Later, it spread to many parts of temperate Europe and USA. The crop was in-troduced in India around 1940 from France for its seed crop.

Area & Production

The crop is cultivated mainly in the states of Punjab (Jallandhar, Gu-daspur and Amritsar districts), Haryana and western Uttar Pradesh (Ladhwa and Saharanpur districts) over an area of about 5000 ha. About 90% of the total produce comes from Punjab.

A one acre plantation of the crop is a highly viable proposition. The cost components of such a model along with the basis for costing are exhibited . The project cost works out to Rs. 125 thousand.

	COST OF PROJECT		
Project Cost: (Unit – One Acre)			(Amount in INR Rs.)

Sl. No.	Component		Proposed Expenditure
1.	Cultivation Expenses		
	(i)	Cost of planting material	1000
	(ii)	Input Cost	5000
	(iii)	Power Cost	2000
	(iv)	Land Preparation	4000
	(v)	Others Farm Operations	4500
		Sub Total	16500
2.	Irrigation		
	(i)	Tubewell / SIP	30000
	(ii)	Electrical Installation etc.	25000
	(iii)	Others, if any	-
		Sub Total	55000
3.	Infrastructure		
	(i)	Pump House	10000
	(ii)	Labourshed	5000
	(iii)	Agriculture Equipments	5000
	(iii)	Others, if any, please specify	
		Sub Total	20000
4.	Land Development		
	(i)	Soil leveling	4000
	(ii)	Digging	-
	(iii)	Fencing	29600
	(iv)	Others, if any, please specify	-
		Sub Total	33600
6.	Land, if newly purchased		-
		Grand Total	125100

<u>Inter-cropping</u>

Labour cost has been put at an average of Rs. 70 per man-day. The actual cost will vary from location to location depending upon minimum wage levels or prevailing wage levels for skilled and unskilled labour.

Recurring Production Cost : Recurring production costs are exhibited in *Annexure III*. The main components are planting material, land preparation, inputs application (FYM, fertilizers, micro-nutrients liming material, plant protection chemicals etc.), power and labour on application of inputs, inter-cultural and other farm operations.

Besides, provision is included for labour for harvesting and transportation charges for the produce to the nearest secondary market.

<u>Returns from the Project</u>: : The value of the produce is estimated at Rs.40000 per annum for the produce of 5 quintal per acre.

PROJECT SYNOPSIS-21

Grape (*Vitis sp.*)

INTRODUCTION

Grape (*Vitis sp.*) belonging to Family Vitaceae is a commercially important fruit crop of India. It is a temperate crop which has got adapted to sub-tropical climate of peninsular India.

OBJECTIVE

The primary objective of this exercise is to support commercial cultivation of grapes by projecting a one acre bankable model project. The high yield of grape is limited to a few vineyards and is not consistent throughout the life - span of the crop. In order to get uniformly high yields with good quality fruit, the basic principles of viticulture needs to percolate down to all the growers.

Peak production during March-April months leading to glut in the market and poor quality of grapes resulting in tremendous post-harvest losses are other problems which limit profits. Growers need to be educated on means of extending harvest over a longer period to get better price for their produce and to minimize market risk.

BACKGROUND

Origin

Grape cultivation is believed to have originated in Armenia near the Caspian Sea in Russia, from where it spread westward to Europe and eastward to Iran and Afghanistan. Grape was introduced in India in 1300 AD by invaders from Iran and Afghanistan.

Area & Production

India is among the first ten countries in the world in the production of grape. The major producers of grape are Italy, France, Spain, USA, Turkey, China and Argentina. This crop occupies fifth position amongst fruit crops in India with a production of 1.21 million tonnes (around 2% of world's production of 57.40 million tonnes) from an area of 0.05 million ha. in 2001-02. The area under grape is 1.2 % of the total area of fruit crops in the country. Production is 2.8% of total fruits produced in the country. About 80% of the production comes from Maharashtra followed by Karnataka and Tamil Nadu.

The cost components of such a model along with the basis for costing are exhibited. A summary is given in the figure below. The project cost works out to Rs. 3.20 lakhs.

Inter-cropping

Labour cost has been put at an average of Rs. 70 per man-day. The actual cost will vary from location to location depending upon minimum wage levels or prevailing wage levels for skilled and unskilled labour. Cost on 'training' system can vary widely depending on the type used. Bower type is the most popular.

Recurring Production Cost : Recurring production costs are exhibited . The main components are farm inputs, (FYM, fertilizers, liming material, plant growth regulators, plant protection chemicals etc.), labour and power mainly for irrigation

Besides, provision is also included for harvesting and packing/ transportation for the produce to the nearest secondary market

Returns from the Project: : The average annual yield from the plantation is estimated at 3 tonnes per acre in year two and 10 tonnes per acre in year three. Valued at Rs.25000 per tonne the total realization works out to Rs.3.25 lakhs per annum.

		COST OF PROJECT	
Sl. No.		Component	Proposed Expenditure
1.		Cultivation Expenses	
	(i)	Cost of planting material	10,000
	(ii)	Manures & fertilizers	9,000
	(iii)	Insecticides & pesticides	4,000
	(iv)	Cost of Labour	8,800
	(v)	Others, if any, (Power)	3600
			35,400
2.		Irrigation	
	(i)	Tube-well/submersible pump	56,000
	(ii)	Cost of Pipeline	-
	(iii)	Others, if any, please specify	-
			56,000
3.		Cost of Drip/Sprinkler	35,000
4.		Infrastructure	
	(i)	Store & pump house	30,000
	(ii)	Labour room	-
	(iii)	Agriculture Equipments	10,000
	(iv)	Others, if any. Please specify (Bower system)	1,20,000
			1,60,000
5.		Land Development	
	(i)	Soil Leveling	4000
	(ii)	Digging	-
	(iii)	Fencing & gates	29600
	(iv)	Others, if any, please specify	-
			33,600
6.		Land, if newly purchased	-
		Grand Total	3,20,000